THE COMPLETE GUIDE TO

DIY RENEWABLE ENERGY

Copyright Page

DIY Renewable Energy

The Complete Guide To

DIY RENEWABLE ENERGY

Does Warren Buffett invest in renewable energy?

By

John King

Expert in Renewable Energy

Introduction

The Rise of Renewable Energy

The world is undergoing a massive shift toward **renewable energy**, driven by the urgent need to combat **climate change**, reduce **carbon emissions**, and secure a sustainable future. As fossil fuel reserves dwindle and energy prices fluctuate, **clean and renewable energy sources**—such as **solar, wind, hydropower, geothermal, and bioenergy**—are becoming the **cheapest, most reliable, and environmentally friendly alternatives**.

Today, renewable energy supplies nearly **30% of the world's electricity**, and with technological advancements, **solar and wind power costs have dropped by over 80% in the last decade**. Nations, businesses, and homeowners alike are investing in **green energy solutions** to **lower energy bills, reduce pollution, and gain energy independence**.

This guide will explore:
✔ **What renewable energy is and why it's important**
✔ **The top renewable energy sources (solar, wind, hydro, etc.)**
✔ **How renewable energy is transforming global power production**
✔ **Why now is the best time to invest in sustainable energy**

By the end, you'll have a **clear understanding** of how renewable energy is reshaping the way we power our homes, businesses, and entire economies—and how you can be a part of the revolution.

Chapter 1: Understanding Renewable Energy and Its Importance

In a world increasingly burdened by the consequences of climate change and resource depletion, renewable energy has emerged as a beacon of hope for a sustainable future. The rising cost of electricity, reliance on fossil fuels, and environmental degradation have motivated individuals to take energy matters into their own hands—literally. DIY renewable energy is more than just a trend; it's a revolutionary movement that empowers everyday people to reduce their carbon footprint, lower energy bills, and achieve a sense of independence.

This chapter sets the foundation for your journey into the world of DIY renewable energy. By the time you finish reading, you'll have a thorough understanding of what renewable energy is, why it's crucial for a sustainable future, and how creating your own systems can make a difference—not just for you, but for the planet.

What is Renewable Energy?

Renewable energy is **energy derived from natural sources** that replenish themselves **faster than they are consumed**. The most common sources of renewable energy include **solar, wind, hydropower, geothermal, and bioenergy**. Unlike **fossil fuels** such as coal, oil, and natural gas—which take **millions of years** to form and emit harmful **greenhouse gases**—renewable energy is **clean, sustainable, and widely available**.

Transitioning from **fossil fuels to renewable energy** is a key step in **reducing carbon emissions**, fighting **climate change**, and ensuring a **sustainable future**. In fact, **renewable energy now accounts for nearly 30% of global electricity production**, and costs have declined **over 80% in the last decade**, making it more **affordable than ever**.

Here are some of the most **common sources of renewable energy**:

Solar Energy

What is Solar Energy?

Solar energy is the most **abundant** of all renewable energy sources, with the sun providing enough power in **one hour** to meet **global energy demand for an entire year**. This energy is captured using **solar panels** and converted into **electricity or heat**.

How Does Solar Energy Work?

✔ **Photovoltaic (PV) Panels**: Convert sunlight directly into electricity.
✔ **Solar Thermal Systems**: Use mirrors to concentrate sunlight and produce heat.

Advantages of Solar Energy

✔ **Zero emissions** – No greenhouse gases during operation.
✔ **Scalable** – Works for small homes or large solar farms.
✔ **Cost-effective** – Solar panel prices have dropped **over 90%** in the last decade.

? **Fact:** The **average lifespan** of a solar panel is **25-30 years**, making it a **long-term investment** for homeowners and businesses.

Wind Energy ?

What is Wind Energy?

Wind energy harnesses the **kinetic energy** of moving air to generate **electricity** using **wind turbines**. This **clean energy source** has been used for centuries and is now a **leading renewable energy technology worldwide**.

Types of Wind Energy Systems

✔ **Onshore Wind Turbines** – Installed on land, producing power for local grids.

✔ **Offshore Wind Farms** – Built in oceans and lakes, where wind speeds are higher.

Advantages of Wind Energy

✔ **Zero fuel cost** – Wind is free and unlimited.
✔ **Scalable** – Small turbines can power homes, while large farms can supply entire cities.
✔ **High efficiency** – Modern wind turbines convert **50-60%** of wind energy into usable power.

? **Fact:** The **largest wind turbine** in the world, located in **Denmark**, generates **15 MW of electricity**, enough to power **20,000 homes**.

Geothermal Energy ?

What is Geothermal Energy?

Geothermal energy comes from **heat stored beneath the Earth's surface**. It can be used for **electricity generation, direct heating, and industrial applications**.

How Does Geothermal Energy Work?

✔ **Geothermal Power Plants** – Use underground heat to generate steam, which spins turbines to produce electricity.
✔ **Geothermal Heat Pumps** – Transfer Earth's heat to provide home heating and cooling.

Advantages of Geothermal Energy

✔ **24/7 power generation** – Not dependent on weather like solar or wind.
✔ **Long lifespan** – Geothermal systems can operate for **100+ years**.
✔ **Small land footprint** – Requires less space compared to solar and wind farms.

? **Fact: Iceland** generates over **90% of its electricity** from geothermal and hydropower, making it one of the most **sustainable energy economies** in the world.

Hydropower ?

What is Hydropower?

Hydropower harnesses the **energy of moving water** to generate **electricity**. It is currently the **largest renewable energy source in the world**, providing about **16% of global electricity**.

Types of Hydropower Systems

✔ **Reservoir Hydropower** – Uses dams to store and release water for electricity production.
✔ **Run-of-River Hydropower** – Generates power without dams, using the natural flow of rivers.
✔ **Pumped Storage** – Stores energy by moving water between two reservoirs.

Advantages of Hydropower

✔ **Reliable and consistent power** – Unlike solar and wind, which depend on weather.
✔ **Multipurpose use** – Dams provide water for **drinking, irrigation, and flood control**.
✔ **Highly efficient** – Converts **90% of water's energy** into electricity.

? **Fact:** The **Three Gorges Dam** in China is the **largest hydroelectric power station** in the world, generating **22,500 MW**—enough to power over **80 million homes**.

Ocean Energy ?

What is Ocean Energy?

Ocean energy uses the **kinetic and thermal energy** of seawater to produce **electricity**. There are several types:

✔ **Tidal Energy** – Captures energy from ocean tides.
✔ **Wave Energy** – Converts wave movement into power.
✔ **Ocean Thermal Energy** – Uses temperature differences between deep and surface waters.

? **Fact:** The **global potential** of ocean energy is estimated at **over 1,000 GW**, enough to power **millions of homes** worldwide.

Bioenergy ?

What is Bioenergy?

Bioenergy is produced from **organic materials** such as **wood, crops, animal waste, and algae**. It can be converted into:
✔ **Solid Biofuels** – Wood pellets, charcoal, and agricultural residues.
✔ **Liquid Biofuels** – Ethanol and biodiesel for transportation.
✔ **Biogas** – Methane produced from organic waste.

Advantages of Bioenergy

✔ **Uses existing waste** – Converts agricultural and food waste into fuel.
✔ **Carbon-neutral** – Plants absorb CO_2, balancing emissions.
✔ **Versatile** – Can replace coal, gasoline, and natural gas.

? **Fact:** Brazil produces **over 40% of its fuel** from bioethanol made from **sugarcane**, reducing reliance on fossil fuels.

The Future of Renewable Energy

The future of **renewable energy** is promising, with **solar, wind, and battery storage costs dropping significantly**. By **2030**, renewables could supply **over 50% of global electricity**, reducing reliance on fossil fuels and cutting **carbon emissions by 70%**.

Key Takeaways:

✔ **Solar and wind are now cheaper than coal and gas.**
✔ **Hydropower remains the largest renewable energy source.**

✓ **Geothermal and ocean energy are emerging technologies.**
✓ **Governments and businesses are investing heavily in renewables.**

? **Transitioning to renewable energy is not just about sustainability—it's the key to an affordable, cleaner, and energy-secure future.**

Renewable energy is power derived from natural sources that are continuously replenished. Unlike fossil fuels such as coal, oil, and natural gas—which take millions of years to form and are finite—renewable energy sources are abundant and sustainable. Examples include:

- **Solar Energy**: Harnessing sunlight to generate electricity or heat.
- **Wind Energy**: Using wind turbines to convert wind currents into electrical power.
- **Hydropower**: Capturing the energy of moving water to produce electricity.
- **Geothermal Energy**: Utilizing heat from the Earth's core for heating and power generation.
- **Biomass and Biogas**: Converting organic waste materials into energy.

The defining feature of renewable energy is that it causes little to no harm to the environment when properly managed, making it a key solution to reducing greenhouse gas emissions and combating global warming.

The Environmental Impact of Fossil Fuels

Before diving into the benefits of renewable energy, it's important to understand the problem we're solving. Fossil fuels, which currently account for the majority of global energy production, come with significant environmental and social consequences:

1. **Carbon Emissions**: Burning fossil fuels releases carbon dioxide (CO_2) into the atmosphere, a primary driver of climate change.
2. **Air Pollution**: The combustion process also produces harmful pollutants like sulfur dioxide (SO_2) and nitrogen oxides (NO_x), which contribute to smog and respiratory issues.
3. **Resource Depletion**: Fossil fuels are non-renewable, meaning they will eventually run out. As supplies dwindle, extraction becomes more destructive and expensive.
4. **Habitat Destruction**: Mining and drilling for coal, oil, and natural gas often destroy ecosystems and displace wildlife.
5. **Water Contamination**: Fracking and oil spills have caused widespread water pollution, harming both human populations and aquatic life.

Transitioning to renewable energy is essential for mitigating these issues and moving toward a cleaner, greener future.

Why Renewable Energy is Crucial for Our Future

The case for renewable energy is compelling, not just for environmental reasons but also for economic and societal benefits.

1. **Combatting Climate Change**
   Renewable energy technologies produce little to no greenhouse gas emissions. By transitioning to renewables, we can significantly reduce the impact of global warming and prevent catastrophic climate events such as rising sea levels, extreme weather, and droughts.

2. **Energy Independence**
   Many countries and individuals are dependent on imported fossil fuels, which leaves them vulnerable to price volatility

and geopolitical conflicts. By producing renewable energy locally, communities can achieve energy security and reduce reliance on external sources.

3. **Lower Energy Costs**
   Although the upfront investment in renewable energy systems can be high, the long-term savings are substantial. Solar panels, wind turbines, and other systems often pay for themselves within a few years, after which the energy they produce is essentially free.
4. **Job Creation**
   The renewable energy sector is labor-intensive, meaning it creates more jobs per unit of energy compared to fossil fuel industries. DIY renewable projects also encourage entrepreneurship and innovation at the community level.
5. **Public Health Benefits**
   Shifting to clean energy reduces air and water pollution, leading to fewer health problems such as asthma, cancer, and heart disease.

The Rise of DIY Renewable Energy

While governments and large corporations play a significant role in the renewable energy transition, individuals are increasingly stepping up to the challenge. DIY renewable energy projects have gained traction for several reasons:

- **Affordability**: DIY solutions are often more cost-effective than hiring professionals or purchasing ready-made systems.
- **Empowerment**: Building your own energy systems provides a sense of independence and accomplishment.
- **Customization**: DIY projects allow you to tailor systems to your specific needs, whether it's powering an off-grid cabin

or reducing energy bills in a suburban home.
- **Community Impact**: DIY projects often inspire neighbors, friends, and local communities to adopt renewable energy, creating a ripple effect of positive change.

Key Benefits of DIY Renewable Energy

1. **Cost Savings**: By cutting out the middleman, you save on labor and markup costs. Many DIY enthusiasts build renewable energy systems for a fraction of the cost of professional installations.
2. **Energy Independence**: Generating your own energy frees you from reliance on utility companies and fluctuating energy prices.
3. **Learning and Skill Development**: DIY projects teach valuable skills, from carpentry and electrical wiring to energy system design.
4. **Environmental Impact**: Every kilowatt-hour of renewable energy you generate reduces demand for fossil fuels.
5. **Resilience**: With a DIY renewable energy system, you're better prepared for power outages, natural disasters, and energy shortages.

Challenges to DIY Renewable Energy (And How to Overcome Them)

While DIY renewable energy is rewarding, it's not without its challenges. Here's how to navigate some common obstacles:

1. **Technical Knowledge**:
    - **Challenge**: Many systems require knowledge of electrical wiring, energy storage, and system design.
    - **Solution**: Start small with simple projects, such as a

solar-powered phone charger, and work your way up to larger systems. There are countless tutorials, online courses, and communities to help you learn.

2. **Upfront Costs**:
   - **Challenge**: The initial investment in materials can be high.
   - **Solution**: Look for secondhand materials, DIY kits, or community grants to lower costs. Many projects can also be completed incrementally.
3. **Permitting and Regulations**:
   - **Challenge**: Some regions have strict laws regarding renewable energy installations.
   - **Solution**: Research local regulations and obtain any necessary permits before starting your project.
4. **Maintenance**:
   - **Challenge**: Renewable energy systems require regular maintenance to operate efficiently.
   - **Solution**: Keep a maintenance log and educate yourself on basic troubleshooting techniques.

Renewable energy is no longer a distant dream; it's an accessible reality, even for DIY enthusiasts. By understanding the importance of renewable energy and the principles behind it, you're taking the first step toward energy independence, environmental stewardship, and financial savings.

This book will guide you through the process of creating your own renewable energy systems, from simple solar setups to advanced wind and hydropower projects. In the next chapter, we'll delve into how to assess your energy needs and create a plan tailored to your goals, budget, and environment.

Your journey to sustainable energy starts here. Let's get started!

Chapter 2: Planning Your DIY Renewable Energy Journey

Embarking on a DIY renewable energy journey is an exciting,

rewarding, and impactful endeavor, but careful planning is the foundation of success. Whether you aim to supplement your energy needs with solar panels or go fully off-grid with a combination of systems, a clear and well-thought-out plan is essential to save time, money, and effort.

This chapter is your step-by-step roadmap for planning your renewable energy projects. From assessing your energy consumption and choosing the right renewable energy source to understanding the costs and space requirements, this chapter will arm you with the knowledge needed to create a blueprint for success. Let's dive in!

Step 1: Conducting a Home Energy Audit

Before jumping into DIY renewable energy projects, you must first understand how much energy your home uses and where you can make improvements to reduce waste. A home energy audit allows you to identify inefficiencies and determine the size and capacity of the renewable energy system you'll need.

What is a Home Energy Audit?
A home energy audit is a thorough assessment of your home's energy consumption. It helps pinpoint where energy is being wasted and suggests ways to improve efficiency. By reducing energy waste, you'll need a smaller, more cost-effective renewable energy system.

How to Perform a DIY Energy Audit:

1. **Check Your Utility Bills**:
    - Review the past 12 months of electricity bills to determine your average monthly and yearly energy usage (measured in kilowatt-hours, or kWh).

- Look for seasonal patterns. For instance, heating in winter or cooling in summer often drives up energy usage.

2. **Inspect Your Insulation and Air Leaks**:
   - Check for gaps around windows, doors, and vents where air may be escaping.
   - Use a thermal imaging camera (or even your hand) to detect drafts or temperature differences.
3. **Examine Your Appliances**:
   - Identify the largest energy consumers in your home (HVAC systems, refrigerators, water heaters, etc.).
   - Consider upgrading to energy-efficient models or implementing smart devices that optimize usage.
4. **Lighting Assessment**:
   - Replace incandescent and halogen bulbs with energy-saving LED lights.
5. **Energy Behavior**:
   - Analyze your habits. Are lights or devices left on unnecessarily? Can you make lifestyle changes to reduce consumption?

Pro Tip: Use free or low-cost tools like smart energy monitors (e.g., Sense or Emporia) to track real-time energy usage.

Step 2: Set Clear Goals for Your Project

Once you have a clear understanding of your energy usage, the next step is to set specific, realistic goals for your DIY renewable energy system. These goals will shape the type and size of the system you build.

Common Goals for DIY Renewable Energy Projects:

1. **Supplemental Energy**:
    - You want to reduce your monthly energy bill by generating part of your energy with renewable sources (e.g., solar panels for daytime use).
    - Example: Powering certain appliances, like a refrigerator or water heater, with renewable energy.
2. **Energy Independence**:
    - The goal is to eliminate reliance on the grid by generating 100% of your own energy. This is particularly useful for remote or off-grid locations.
3. **Emergency Backup Power**:
    - You want a renewable energy system to provide power during outages. This might include a solar generator or small battery system.
4. **Environmental Impact**:
    - Your primary motivation may be to reduce your carbon footprint and live sustainably.
5. **Scalability**:
    - Start small and build over time, creating a system that can grow as your energy needs or budget expand.

Step 3: Choose the Right Renewable Energy System

Selecting the right renewable energy source depends on your location, budget, and energy needs. Below, we'll break down the most common DIY renewable energy systems and how to choose the one(s) that best fit your goals.

1. Solar Power Systems

- **Best For**: Locations with abundant sunlight.
- **Advantages**:
    - Solar panels are relatively easy to install and maintain.
    - Costs have dropped significantly in recent years.
- **Considerations**:
    - Solar requires roof or ground space with direct sunlight exposure.
    - Energy production can drop on cloudy days or at night.
- **Pro Tip**: Use tools like PVWatts (by NREL) to estimate solar energy potential for your location.

2. Wind Turbines

- **Best For**: Locations with consistent wind speeds of at least 10-12 mph.
- **Advantages**:
    - Produces energy day and night as long as there's wind.
    - Can be combined with solar for a hybrid system.
- **Considerations**:
    - Wind turbines are more complex to build and require open land.
    - They can be noisy and may face local zoning restrictions.

3. Hydropower Systems

- **Best For**: Properties with access to a flowing water source (stream or river).
- **Advantages**:
    - Produces consistent energy as long as water flow remains steady.

    - More efficient than wind or solar in terms of energy output.
- **Considerations**:
    - Requires a reliable and accessible water source.
    - Installation can be more complex and require permits.

4. Geothermal Energy Systems

- **Best For**: Locations with moderate-to-high underground heat availability.
- **Advantages**:
    - Provides heating, cooling, and electricity with low ongoing costs.
- **Considerations**:
    - Initial installation can be expensive and labor-intensive.

5. Biogas Systems

- **Best For**: Properties with organic waste (e.g., farms or homes with compostable materials).
- **Advantages**:
    - Converts waste into usable energy while reducing landfill contributions.
    - Systems can be simple to build and maintain.
- **Considerations**:
    - Energy output is lower compared to solar or wind.

Step 4: Budgeting for Your DIY Project

Understanding the costs involved is critical to ensuring the success of your project. DIY renewable energy systems can range from small, low-cost projects to full-scale installations requiring significant investment.

Cost Breakdown:

1. **Upfront Costs**:
    - Solar panels, wind turbine components, batteries, inverters, wiring, mounting equipment, and tools.
2. **Maintenance Costs**:
    - Replaceable components like batteries (every 5-10 years).
3. **Permits and Inspections**:
    - Some projects require building permits and inspection fees.

Pro Tips for Budgeting:

- Start small and scale up as funds become available.
- Look for secondhand materials or DIY kits to save money.
- Explore government rebates, tax credits, and grants for renewable energy projects in your region.

Step 5: Assess Space and Location Requirements

Each renewable energy system has unique space requirements. Consider the following:

- **Solar Panels**:
    - Requires a roof or ground space with direct sunlight exposure for most of the day.
- **Wind Turbines**:
    - Needs open land with no obstructions (trees, buildings) to ensure wind flow.
    - Small turbines can be mounted on rooftops, but larger ones require open areas.
- **Hydropower Systems**:
    - Must be installed near a water source with consistent

flow.

- **Battery Storage**:
  - Requires a safe, well-ventilated space (like a garage or shed) for batteries and inverters.

Step 6: Create a Timeline and Project Plan

Renewable energy projects can take anywhere from a few days to several months, depending on the size and complexity. Break your project into manageable phases:

1. **Research**: Gather resources, materials, and knowledge.
2. **Design**: Plan your system layout, calculate energy needs, and acquire permits.
3. **Build**: Assemble and install components.
4. **Test and Optimize**: Ensure everything is functioning properly before scaling up.

Planning is the most critical step in your DIY renewable energy journey. By conducting a home energy audit, setting clear goals, selecting the right system, budgeting wisely, and assessing space requirements, you'll set yourself up for success. The effort you put into planning now will save you time, money, and headaches later.

In the next chapter, we'll take a closer look at solar power systems—the most popular DIY renewable energy option—and guide you through designing and building your own. Let's bring your renewable energy plan to life!

Let me know if you'd like additional sections or expansions! ?

Chapter 3: Harnessing the Power of the Sun – DIY Solar Energy

Solar energy is one of the most accessible, cost-effective, and scalable renewable energy sources available today. With the cost of solar panels dropping dramatically over the last decade and efficiency improvements making DIY installations more effective than ever, solar power is an attractive option for homeowners, off-grid enthusiasts, and eco-conscious individuals looking to reduce their reliance on fossil fuels.

This chapter will serve as a **comprehensive, step-by-step guide to designing, building, and maintaining a DIY solar energy system**. Whether you want to power your entire home, supplement your electricity needs, or create a small off-grid system, you'll find everything you need to get started.

Section 1: Understanding How Solar Power Works

Before jumping into designing and installing your DIY solar system, it's important to understand the basic mechanics of how solar energy is converted into electricity.

1.1 How Solar Panels Generate Power

Solar panels work by using **photovoltaic (PV) cells**, which convert sunlight into electricity through the **photovoltaic effect**. Here's how it works:

1. Sunlight hits the PV cells, exciting electrons within the semiconductor material (usually silicon).
2. This movement of electrons generates direct current (DC) electricity.
3. An **inverter** converts DC electricity into alternating current (AC), which is what most household appliances use.
4. The electricity is either used immediately, stored in a battery

system, or sent back to the power grid (if you have a grid-tied system).

1.2 Key Components of a DIY Solar System

A solar energy system consists of several important components:

Component	Function
Solar Panels	Capture sunlight and convert it into DC electricity.
Charge Controller	Regulates voltage and current to prevent battery overcharging. (Used in off-grid systems.)
Battery Storage (Optional)	Stores energy for later use (useful for off-grid or backup power).
Inverter	Converts DC electricity into AC for household appliances.
Mounting System	Holds solar panels securely on a roof or ground structure.
Wiring & Connectors	Transfers electricity between components.

Section 2: Designing Your DIY Solar Energy System

Proper planning ensures that your solar energy system is efficient, reliable, and cost-effective.

2.1 Calculating Your Energy Needs

To determine how many solar panels you need, follow these steps:

Step 1: Find Your Daily Energy Consumption

- Look at your electricity bill and find the monthly kilowatt-hour (kWh) usage.

- Divide by 30 to get your daily usage.
- Example: If your household consumes **900 kWh per month**, then:
  - **900 ÷ 30 = 30 kWh per day**

Step 2: Estimate Solar Panel Output in Your Location

- Solar panel output varies by location.
- Use a tool like **PVWatts Calculator (by NREL)** to estimate sunlight availability.
- A **300-watt solar panel** produces approximately **1.2 kWh per day** in a sunny area.

Step 3: Determine How Many Panels You Need

- If you need **30 kWh per day** and each panel produces **1.2 kWh/day**, then:
  - **30 ÷ 1.2 = 25 solar panels**

2.2 Choosing the Right Solar Panels

There are three main types of solar panels:

Panel Type	Efficiency	Cost	Best For
Monocrystalline	High (18-22%)	Expensive	Small spaces needing maximum efficiency.
Polycrystalline	Medium (15-18%)	Moderate	Budget-conscious DIYers with moderate space.
Thin-Film	Low (10-12%)	Cheap	Portable applications or large, inexpensive installations.

For most **home-based DIY solar projects, monocrystalline**

panels are the best choice due to their efficiency and durability.

2.3 Grid-Tied vs. Off-Grid Solar Systems

Your system can either be **grid-tied** (connected to your local electric utility) or **off-grid** (completely independent).

System Type	Pros	Cons
Grid-Tied	Can sell excess electricity to the utility company (net metering), reduces bills.	No backup power during outages.
Off-Grid	Complete energy independence, works in remote areas.	Requires expensive battery storage, no utility backup.

Section 3: Step-by-Step Guide to Installing Your DIY Solar System

Now that you've designed your system, let's go through the installation process step by step.

3.1 Tools and Materials Needed

- **Solar panels**
- **Inverter**
- **Battery bank (for off-grid systems)**
- **Charge controller**
- **Mounting hardware**
- **Wiring and connectors**
- **Drill, wrench, and basic hand tools**

3.2 Installing the Solar Panels

Step 1: Choose the Best Location

- Roof-mounted panels: Best for homes with unshaded, south-facing roofs.
- Ground-mounted panels: Ideal for properties with ample open land.

Step 2: Install the Mounting System

- Use rails or brackets to securely attach panels to your roof or a ground-based rack.
- Ensure panels are tilted at an optimal angle (usually **30-40 degrees** for maximum sunlight exposure).

Step 3: Secure the Solar Panels

- Bolt the panels onto the mounting structure.
- Leave space between panels for air circulation and cooling.

3.3 Wiring and Connecting the System

Step 1: Connect the Panels

- Use **MC4 connectors** to wire panels in **series or parallel**, depending on your voltage needs.

Step 2: Connect to Charge Controller (Off-Grid Systems Only)

- Regulates energy flow to prevent battery damage.

Step 3: Connect the Inverter

- Converts DC power from panels into usable AC power.

Step 4: Connect to Your Home's Electrical System (Grid-Tied Systems Only)

- Requires a **net meter** and proper permits.
- Consult a licensed electrician if unsure.

Section 4: Maintaining and Troubleshooting Your DIY Solar System

4.1 Regular Maintenance Tips

- **Clean Panels Every 6 Months**: Dust and debris reduce efficiency.
- **Check Wiring and Connections**: Look for corrosion or loose fittings.
- **Monitor Energy Production**: Use a solar monitoring app to track performance.

4.2 Troubleshooting Common Issues

Issue	Possible Cause	Solution
Low Energy Output	Shading, dirty panels, inverter issues	Clean panels, check for obstructions.
Battery Not Charging	Faulty charge controller, old batteries	Test and replace defective components.
Inverter Not Functioning	Loose connections, overload	Check wiring and reset the inverter.

Building your own DIY solar energy system is one of the most empowering steps toward energy independence and sustainability.

By carefully planning, choosing the right components, and following proper installation techniques, you can harness the sun's power to reduce your electricity costs, decrease your environmental footprint, and take control of your energy future.

In **Chapter 4**, we will explore **DIY wind turbines**, another excellent renewable energy source that can complement your solar setup. Get ready to capture the power of the wind!

Chapter 4: Catching the Wind – DIY Wind Turbines

Wind energy has been harnessed for centuries, from ancient windmills grinding grain to modern wind turbines generating electricity on a massive scale. But you don't need a commercial wind farm to benefit from this powerful and renewable resource—**DIY wind turbines** are an excellent way to generate electricity for your home, cabin, or off-grid setup.

In this chapter, we will explore everything you need to know about building and installing a DIY wind turbine, including:

- How wind energy works
- Assessing your location and wind potential
- Choosing the right wind turbine type
- Step-by-step instructions for building and installing your system
- Maintenance, troubleshooting, and safety considerations

By the end of this chapter, you'll be equipped with the knowledge and confidence to build your own **DIY wind power system**, reducing your reliance on fossil fuels and lowering your electricity costs.

Section 1: Understanding How Wind Power Works

1.1 How Wind Turbines Generate Electricity

Wind turbines convert kinetic energy from the wind into mechanical energy, which is then transformed into electrical power. Here's how it works:

1. **Wind Blades Capture Energy**: Wind pushes against the turbine's blades, causing them to rotate.
2. **Rotor and Shaft Convert Motion**: The blades are connected to a rotor, which spins a shaft inside the turbine.

3. **Generator Produces Electricity**: The spinning shaft turns a generator, producing **direct current (DC) electricity**.
4. **Inverter Converts to AC Power**: If you're using the power in your home, an **inverter** converts DC power into **alternating current (AC)**, which is used by household appliances.
5. **Battery Storage (Optional)**: If you want to store power for later use, you can connect the system to a **battery bank**.

1.2 Key Components of a DIY Wind Turbine

A DIY wind turbine consists of several essential parts:

Component	Function
Blades	Capture wind energy and rotate.
Rotor & Hub	Connects the blades to the turbine's main shaft.
Shaft	Transfers mechanical motion from the rotor to the generator.
Generator (Alternator)	Converts mechanical energy into electricity.
Tail Vane	Keeps the turbine facing into the wind.
Tower/Pole	Elevates the turbine to catch more wind.
Charge Controller (Optional)	Regulates power going into batteries (for off-grid systems).
Battery Bank (Optional)	Stores electricity for later use.
Inverter	Converts DC power to AC for household use.

Section 2: Assessing Your Wind Energy Potential

2.1 Do You Have Enough Wind?

Not every location is ideal for wind power. You need **consistent, strong winds** for a wind turbine to be effective.

How to Measure Wind Speed

- **Use an Anemometer**: A handheld or mounted anemometer will measure wind speed at your site.
- **Check Online Wind Maps**: Websites like **Global Wind Atlas** or **NOAA Wind Data** provide wind speed estimates for your area.
- **Look for Wind Indicators**: Trees bent in one direction, flag movement, and local weather patterns can give you clues about average wind speeds.

Minimum Wind Speed Requirement:

- Small DIY wind turbines require at least **8-10 mph (13-16 km/h)** for decent performance.
- **Optimal wind speeds** are **12-25 mph (19-40 km/h)**.
- If your wind speeds are below **6 mph (10 km/h)**, wind power may not be practical.

2.2 Choosing the Best Location

To get the most from your wind turbine, place it:
✓ **At least 30 feet above obstructions** (trees, buildings)
✓ In an **open field or elevated area** for maximum wind exposure
✓ **Away from wind turbulence** caused by structures

Section 3: Choosing the Right Wind Turbine

3.1 Horizontal vs. Vertical Wind Turbines

There are two main types of wind turbines:

Type	Pros	Cons
Horizontal-Axis Wind Turbine (HAWT)	More efficient, produces more power, proven technology.	Needs steady wind, complex construction, requires a high tower.
Vertical-Axis Wind Turbine (VAWT)	Works in low-wind areas, easy to build, can be mounted on rooftops.	Less efficient, requires more maintenance.

? **Best Choice for DIY**: Most DIYers build **Horizontal-Axis Wind Turbines (HAWTs)** due to their higher efficiency and power output.

Section 4: Step-by-Step Guide to Building Your DIY Wind Turbine

4.1 Materials and Tools Needed

Materials:

- **Blades**: Made from PVC pipes, wood, or metal.
- **Generator**: A **DC motor** (from old treadmills or appliances) works well.
- **Tower**: Steel or PVC pipe, at least 10-20 feet high.
- **Tail Vane**: Made from sheet metal or wood.
- **Wiring**: Copper wire for electrical connections.
- **Charge Controller & Batteries (for off-grid use)**.

Tools:

- Power drill
- Angle grinder
- Wrench set
- Welding tools (if using metal blades)
- Multimeter (for voltage testing)

4.2 Assembling the Wind Turbine

Step 1: Build the Rotor and Blades

- Cut **3-5 blades** from **PVC, wood, or aluminum**.
- Shape the blades aerodynamically for efficiency.
- Attach blades to a **hub** (a metal or wooden disc).

Step 2: Attach the Generator

- Mount a **permanent magnet DC motor** to the rotor shaft.
- Secure it with a bracket or welded frame.

Step 3: Install thc Tail Vane

- The tail keeps the turbine aligned with the wind.
- Cut **lightweight metal or plywood** and attach it to the frame.

Step 4: Construct the Tower

- Use **a steel pipe or strong wood** (10-20 feet high).
- Secure it with **guy wires** for stability.

Step 5: Wiring the System

- **Grid-Tied System**: Connect to an inverter and feed power into your home.
- **Off-Grid System**: Use a **charge controller** to regulate energy storage in batteries.

Section 5: Installing and Maintaining Your Wind Turbine

5.1 Installation Tips

- Install your turbine on a sturdy **tilting tower** for easy maintenance.
- Ensure **wiring is weatherproof**.
- Mount **batteries and electronics indoors** to protect them.

5.2 Maintenance & Troubleshooting

✅**Inspect Blades Every 3 Months**: Check for cracks or warping.
✅**Tighten Bolts & Fasteners**: Wind vibration can loosen screws.
✅**Lubricate Bearings**: Keeps the system running smoothly.
✅**Check Electrical Output**: Use a **multimeter** to measure voltage output.

DIY wind turbines are a **powerful and cost-effective way to generate renewable energy**. By assessing your wind potential, choosing the right turbine design, and following this step-by-step guide, you can successfully harness wind power for your home or off-grid setup.

In **Chapter 5**, we'll explore **DIY hydropower systems**—another reliable and efficient renewable energy source. Stay tuned!

Let me know if you'd like more **diagrams, cost breakdowns, or troubleshooting sections**!

Chapter 5: DIY Hydropower – Generating Energy from Water

Hydropower is one of the oldest and most reliable sources of renewable energy. Long before the advent of electricity, civilizations harnessed the energy of flowing water to grind grain, pump water, and perform other essential tasks. Today, **micro-hydropower systems** allow individuals to generate their own electricity from small streams, rivers, or even backyard water features.

Hydropower has several advantages over **solar and wind energy**:
✔ **Continuous power generation** – Unlike solar (which only works in daylight) or wind (which depends on wind speeds), hydropower can generate electricity **24/7**, as long as water is flowing.
✔ **High efficiency** – Hydropower systems can convert **60-90% of water energy** into usable electricity, making them one of the most efficient renewable energy sources.
✔ **Reliable and low maintenance** – With fewer moving parts than wind turbines, small hydropower systems require **minimal upkeep**.

This chapter will guide you through **designing, building, and installing your own DIY hydropower system**, covering:

- How hydropower works
- Assessing your site for feasibility
- Choosing the right hydropower system
- Step-by-step construction
- Maintenance and troubleshooting

Section 1: How Hydropower Works

1.1 The Science Behind Hydropower

Hydropower works by **harnessing the kinetic energy** of moving water and converting it into mechanical energy, which then drives a generator to produce electricity. The key principles of hydropower include:

- **Flow (Q)**: The volume of water moving per second (measured in liters per second or cubic feet per second).
- **Head (H)**: The vertical drop of the water from the intake to the turbine (measured in meters or feet).
- **Power Output (P)**: The amount of electrical energy generated, determined by the equation:

P=Flow(Q)×Head(H)×9.81×EfficiencyP = \text{Flow} (Q) \times \text{Head} (H) \times 9.81 \times \text{Efficiency}P=Flow(Q)×Head(H)×9.81×Efficiency

Where **9.81** is the gravitational acceleration constant, and efficiency depends on the turbine type (typically 60-90%).

4.2 Key Components of a Hydropower System

A **micro-hydro system** consists of several key parts:

Component	Function
Intake System	Collccts water from the source and directs it to the pipeline.
Penstock (Pipeline)	Carries water from the intake to the turbine.
Turbine	Converts the energy of moving water into rotational motion.
Generator (Alternator)	Converts mechanical energy from the turbine into electricity.
Control System	Regulates voltage and power output.
Battery Bank (Optional)	Stores energy for later use.

Section 2: Assessing Your Hydropower Potential

Before building a DIY hydropower system, you must **evaluate your site** to ensure sufficient water flow and head.

2.1 Measuring Water Flow (Q)

Flow rate is the volume of water passing through the system per second. There are several ways to measure it:

- **Bucket Method** (For Small Streams)
    1. Use a container of known volume (e.g., a **10-liter bucket**).
    2. Time how long it takes to fill up completely from the water source.
    3. Calculate the flow rate using:

  Q=Bucket Volume (liters)Time (seconds)Q = \frac{\text{Bucket Volume (liters)}}{\text{Time (seconds)}}Q=Time (seconds)Bucket Volume (liters)
- **Float Method** (For Larger Streams)
    1. Mark a 10-meter section of the stream.
    2. Drop a floating object (e.g., a small wood block) and measure the time it takes to travel 10 meters.
    3. Multiply by the stream's width and depth to estimate total flow.

2.2 Measuring Head (H)

Head is the vertical height difference between the water intake and the turbine. You can measure it using:
✔ **Surveyor's Level** – Best for long distances.
✔ **Smartphone Altimeter Apps** – Useful for quick estimates.
✔ **Simple Hose Level** – A clear hose filled with water can help

measure small height differences.

Example Calculation

- **Flow (Q):** 10 liters per second
- **Head (H):** 3 meters
- **Power Output:**

P=10×3×9.81×0.7=206 wattsP = 10 \times 3 \times 9.81 \times 0.7 = 206 \text{ watts}P=10×3×9.81×0.7=206 watts

With a **higher head**, you need **less flow** for the same power output. **Lower head** requires a **higher flow rate**.

Section 3: Choosing the Right Hydropower System

There are different types of hydropower turbines, each suited for different conditions:

Turbine Type	Best For	Efficiency
Pelton Wheel	High-head, low-flow sites	80-90%
Turgo Turbine	Medium-head, moderate-flow sites	70-85%
Crossflow Turbine	Medium- to low-head sites	60-75%
Propeller Turbine	Low-head, high-flow sites	50-70%

For **DIY projects, Pelton wheels and crossflow turbines** are the most practical options.

Section 4: Step-by-Step Guide to Building a DIY Hydropower System

4.1 Tools and Materials Needed

✅PVC or steel pipe (for the penstock)
✅Plastic or metal wheel (for the turbine)
✅Alternator (from a car or old generator)
✅Voltage regulator
✅Concrete and rebar (for securing components)

4.2 Building the System

Step 1: Construct the Water Intake

- Build a **small dam or weir** to collect water.
- Install a **screen filter** to remove debris.

Step 2: Install the Penstock (Pipeline)

- Use **PVC, HDPE, or steel pipe**.
- Ensure a **steady downhill slope** to maintain pressure.

Step 3: Build the Turbine Housing

- Mount the turbine at the lowest point of the system.
- Ensure **water exits properly** to prevent flooding.

Step 4: Connect the Generator

- Attach a **DC generator or alternator** to the turbine shaft.
- Use **a belt or direct coupling** for efficient power transfer.

Step 5: Wire the Electrical System

- **Grid-Tied System:** Connect to an **inverter**.
- **Off-Grid System:** Use a **charge controller and battery bank**.

Section 5: Maintenance and Troubleshooting

5.1 Regular Maintenance Checklist

✔ **Clean the intake filter** weekly.
✔ **Inspect pipes and fittings** for leaks.
✔ **Lubricate bearings** every few months.
✔ **Check voltage output** to ensure stable power.

5.2 Troubleshooting Common Issues

Problem	Possible Cause	Solution
Low Power Output	Blocked intake, low water flow	Clean intake, check seasonal changes.
No Power	Generator failure	Test with a multimeter, replace faulty components.
Water Leaks	Loose pipe connections	Tighten fittings, seal leaks.

Building your own **DIY hydropower system** is one of the most efficient and reliable ways to generate **renewable electricity year-round**. By following this guide, you can design, construct, and maintain a **self-sustaining power source** that outperforms solar and wind in the right conditions.

Chapter 6: Energy Storage Solutions for DIY Renewable Systems

Renewable energy sources like **solar, wind, and hydropower** offer an incredible opportunity for energy independence, but one major challenge remains: **energy availability fluctuates**.

- Solar panels **only generate power during the day**,
- Wind turbines **only produce electricity when the wind is blowing**,
- Hydropower systems **may experience seasonal changes in water flow**.

To ensure a **steady, reliable** power supply, **energy storage** is essential. **Batteries, inverters, and charge controllers** allow you to store excess power for later use, providing backup energy during nights, cloudy days, or low-wind periods.

In this chapter, you'll learn:
✅The different types of **battery storage systems**
✅How to **calculate the battery capacity** you need
✅The role of **charge controllers and inverters**
✅How to **design, install, and maintain a DIY battery bank**
✅The future of **advanced energy storage technologies**

By the end of this chapter, you'll have a **fully functional energy storage plan** for your DIY renewable system, ensuring **reliable, continuous power** for your home, cabin, or off-grid project.

Section 1: Why Energy Storage is Essential

1.1 The Problem of Energy Intermittency

Without energy storage, **renewable power systems face major challenges**:

Energy Source	Issue Without Storage
Solar Panels	Power is only available during the day. No electricity at night.
Wind Turbines	Power is inconsistent, fluctuating with wind speeds.
Hydropower	Water levels can vary, reducing electricity production.

1.2 The Role of Energy Storage

Batteries **store excess power** and release it when needed, making renewable energy practical for daily use.

✔ **Grid-tied systems** use energy storage to reduce reliance on utility companies and lower electricity bills.
✔ **Off-grid systems** depend on batteries to provide power when no generation is available.
✔ **Hybrid systems** combine renewables with battery storage and backup generators for 24/7 power.

Section 2: Choosing the Right Battery for Your DIY System

Not all batteries are created equal. When selecting a **battery type**, consider:

- **Capacity** (how much energy it can store)
- **Lifespan** (how many charge cycles it lasts)
- **Cost** (upfront and long-term value)
- **Efficiency** (energy loss during charging/discharging)
- **Maintenance requirements**

2.1 Types of Batteries for Renewable Energy Storage

Battery Type	Pros	Cons	Lifespan
Lead-Acid (Flooded)	Low cost, widely available	Requires maintenance, shorter lifespan	3-5 years
AGM (Absorbent Glass Mat)	Maintenance-free, safer than flooded lead-acid	More expensive than standard lead-acid	5-7 years
Gel Batteries	Spill-proof, long-lasting	Cannot handle high charge/discharge rates well	5-7 years
Lithium-Ion (LiFePO4)	High efficiency, long lifespan, maintenance-free	Expensive upfront, but best long-term investment	10-15 years
Saltwater Batteries	Eco-friendly, long lifespan	Limited availability, high cost	10-15 years

2.2 Best Choice for DIY Renewable Systems

? **For budget-conscious DIYers**: AGM or Gel batteries
? **For long-term reliability**: **Lithium-Ion (LiFePO4)** is the best choice despite the higher initial cost.
? **For sustainable projects**: Saltwater batteries are eco-friendly and last longer.

Section 3: How to Calculate Your Battery Storage Needs

To size your **DIY battery bank**, follow these steps:

3.1 Step 1: Determine Daily Energy Consumption

Find your average **daily energy use** in **kilowatt-hours (kWh)**.

- Example: If your household uses **30 kWh per day**, you need enough battery storage to cover this demand.

3.2 Step 2: Factor in Days of Autonomy

How many days should your battery bank provide power **without recharging**?

- If you want **3 days of backup**, multiply:

30kWh×3=90kWh30 kWh \times 3 = 90 kWh30kWh×3=90kWh

3.3 Step 3: Convert to Battery Capacity

Batteries are rated in **amp-hours (Ah)**, not kWh. Use this formula:

Battery Capacity (Ah)=Total Energy (kWh)×1000Battery Voltage (V)\text{Battery Capacity (Ah)} = \frac{\text{Total Energy (kWh)} \times 1000}{\text{Battery Voltage (V)}}Battery Capacity (Ah)=Battery Voltage (V)Total Energy (kWh)×1000

For a **48V lithium battery bank**:

90,00048=1,875Ah\frac{90,000}{48} = 1,875 Ah4890,000=1,875Ah

This means you need **1,875 Ah of battery capacity at 48V** to sustain power for 3 days.

Section 4: Charge Controllers – Protecting Your Battery Bank

4.1 What is a Charge Controller?

A **charge controller** regulates energy flow **from solar panels or wind turbines to batteries**, preventing:
✔ **Overcharging** (which damages batteries)
✔ **Overdischarging** (which shortens lifespan)
✔ **Voltage fluctuations**

4.2 Types of Charge Controllers

Type	Pros	Cons
PWM (Pulse Width Modulation)	Low cost, simple design	Less efficient, works best with lead-acid batteries
MPPT (Maximum Power Point Tracking)	30% more efficient, maximizes energy use	Higher cost, requires proper configuration

Best Choice: **MPPT charge controllers** maximize efficiency, especially for **solar** and **wind** power systems.

Section 5: Inverters – Converting DC to Usable AC Power

Most household appliances run on **AC power (120V or 240V)**, but **batteries store DC power**. An **inverter** is required to convert:

✔ **DC (direct current) → AC (alternating current)**

5.1 Choosing the Right Inverter

Inverter Type	Best For
Modified Sine	Simple loads (lights, fans)

Inverter Type	Best For
Wave	
Pure Sine Wave	All household appliances, sensitive electronics

? **For most DIY systems, a pure sine wave inverter is recommended.**

Section 6: Installing and Maintaining Your Battery Bank

6.1 DIY Battery Installation Steps

✔ **Choose a ventilated, temperature-controlled area**
✔ **Wire batteries in series (for higher voltage) or parallel (for higher capacity)**
✔ **Connect batteries to charge controller and inverter**

6.2 Maintenance Checklist

✅ **Lead-Acid Batteries**: Check water levels monthly.
✅ **Lithium Batteries**: Keep at 20-80% charge for longevity.
✅ **Check Voltage Levels**: Avoid deep discharges below 50%.

Section 7: The Future of Energy Storage

? **Solid-State Batteries** – Safer, longer lifespan, higher energy density
? **Hydrogen Fuel Cells** – Large-scale storage potential
? **Supercapacitors** – Fast-charging, high-power output

With advances in battery technology, **energy storage is becoming cheaper, more efficient, and more sustainable**.

A **properly designed battery storage system** ensures that your **DIY renewable energy setup provides continuous, reliable**

power. By choosing the right batteries, charge controllers, and inverters, you can build a system that suits your energy needs while maximizing efficiency and longevity.

Chapter 7: Off-Grid Living with Renewable Energy

Imagine a life where you are completely independent of utility companies, free from power outages, and in full control of your energy needs. **Off-grid living** is the ultimate goal for many DIY renewable energy enthusiasts, offering sustainability, cost savings, and energy security. Whether you're planning to go fully off-grid or simply reduce your reliance on the electrical grid, this chapter will provide a **detailed guide to achieving energy independence**.

In this chapter, you'll learn:
✔ What off-grid living means and its benefits
✔ How to design an off-grid **renewable energy system**
✔ Essential **water, heating, and cooling solutions**
✔ The best ways to store and manage **electricity**
✔ Off-grid communication, security, and maintenance tips

By the end, you'll have the **knowledge and tools** to create a self-sufficient, off-grid lifestyle powered by renewable energy.

Section 1: What is Off-Grid Living?

1.1 Definition of Off-Grid Living

Off-grid living means living **without reliance on public utilities**, particularly the electrical grid. Instead, power is generated **on-site** using renewable energy systems like **solar panels, wind turbines, or hydropower**.

Some off-grid setups are fully independent, while others remain partially connected to municipal water or waste services.

1.2 Benefits of Off-Grid Living

✔ **Energy Independence** – No more electricity bills or reliance on power companies.
✔ **Sustainability** – Reduced carbon footprint with **100%**

renewable energy use.
✔ **Cost Savings** – After the initial investment, energy is essentially **free**.
✔ **Resilience** – Protected from power outages, fuel shortages, and grid failures.
✔ **Freedom** – The ability to live in **remote locations** where grid power is unavailable.

Section 2: Designing Your Off-Grid Renewable Energy System

Going off-grid requires a **combination of renewable energy sources** and **efficient energy storage** to ensure 24/7 power availability.

2.1 Choosing the Right Energy Source

The best off-grid power system depends on **your location** and **energy needs**.

Energy Source	Best For	Pros	Cons
Solar Power	Most locations with good sunlight	Reliable, scalable, low maintenance	No power at night, reduced efficiency in winter
Wind Power	Windy areas (coastal, plains)	Works day & night, complements solar	Requires space and high towers
Hydropower	Locations with a steady stream or river	Constant power generation, highly efficient	Requires flowing water source, legal restrictions
Biogas/Gen	Emergency	Provides power	Requires fuel,

Energy Source	Best For	Pros	Cons
erator	backup	when renewables are low	emits CO_2

? **Pro Tip**: A **hybrid system** (solar + wind + battery storage) ensures **power reliability** in all conditions.

2.2 Estimating Your Energy Needs

1. **List all electrical devices** (lights, appliances, electronics).
2. **Calculate total watt-hours per day**.
3. **Size your battery bank** (See Chapter 6 for details).
4. **Choose an inverter that matches your total power load**.

Example:

- If your daily energy use is **15 kWh**, you'll need:
    - **4-5 kW solar panels**
    - **20 kWh battery storage** for nighttime use
    - A **5 kW inverter** to handle peak loads

Section 3: Off-Grid Water Systems

Water is **as important as electricity** for off-grid survival. Here's how to secure, store, and purify water without relying on municipal sources.

3.1 Sourcing Off-Grid Water

- **Well Water** – Ideal if groundwater is accessible. Requires a **solar or wind-powered pump**.
- **Rainwater Harvesting** – Collecting and storing rainwater in large tanks.
- **Stream or River** – Reliable but may require filtration and

legal permissions.

3.2 DIY Water Filtration

Off-grid homes need **filtration and purification** systems to ensure clean drinking water.

✔ **Sediment Filters** – Removes dirt, sand, and debris.
✔ **Carbon Filters** – Eliminates chlorine, bad taste, and some chemicals.
✔ **UV Purifiers** – Kills bacteria and viruses.
✔ **Reverse Osmosis** – Best for removing heavy metals and fluoride.

? **Pro Tip**: Install a **solar-powered water pump** to reduce electricity use.

Section 4: Off-Grid Heating and Cooling Solutions

4.1 Heating Your Home Off-Grid

- **Passive Solar Design** – Use **south-facing windows** to capture heat.
- **Rocket Mass Heaters** – Burns wood efficiently, providing **long-lasting warmth**.
- **Solar Water Heaters** – Heat water using **solar thermal collectors**.
- **Wood Stoves** – A classic, reliable heating method.

4.2 Cooling an Off-Grid Home

- **Earth Tubes** – Underground pipes that **naturally cool** incoming air.
- **Solar-Powered Fans** – Provides ventilation without grid

power.
- **Thermal Mass** – Stone and concrete absorb heat during the day and release it at night.

Section 5: Off-Grid Refrigeration and Cooking

5.1 Refrigeration Options

✓ **Solar-Powered Refrigerators** – Runs on **12V/24V DC**, ideal for off-grid use.
✓ **Propane Refrigerators** – Uses **liquid propane**, common in remote cabins.
✓ **Root Cellars** – Underground storage for fruits, vegetables, and dairy.

5.2 Cooking Without Grid Power

? **Rocket Stoves** – Highly efficient wood-burning stoves.
☀ **Solar Ovens** – Uses the sun to bake, boil, and fry food.
? **Biogas Digesters** – Converts organic waste into cooking gas.

Section 6: Off-Grid Communication and Security

6.1 Staying Connected

- **Satellite Internet (Starlink, HughesNet)** – High-speed, works anywhere.
- **Ham Radio** – Essential for emergencies and long-range communication.
- **Cell Signal Boosters** – Improves weak mobile signals in remote areas.

6.2 Security Measures

- **Solar-Powered Cameras** – Motion-activated surveillance cameras.
- **Perimeter Fencing** – Keeps livestock in and intruders out.
- **Dogs & Livestock Guardians** – Natural alarms against trespassers.

Section 7: Waste Management and Composting

7.1 Off-Grid Septic Systems

- **Composting Toilets** – Turns waste into **fertilizer** with no plumbing required.
- **Septic Tanks** – Requires occasional pumping but works like a standard system.
- **Greywater Recycling** – Filters used water for irrigation.

7.2 Composting and Recycling

✔ **Vermicomposting (Worm Bins)** – Converts kitchen scraps into soil nutrients.
✔ **Biogas Digesters** – Creates methane gas for cooking.
✔ **Recycling Centers** – Find local drop-off points for cans, plastics, and paper.

Living off-grid is a **challenging but rewarding** lifestyle that requires planning, creativity, and self-sufficiency. By using **renewable energy**, efficient water systems, and sustainable practices, you can **live comfortably without reliance on the grid**.

Chapter 8: Advanced DIY Renewable Energy Projects – Microgrids, Smart Automation, and Hybrid Systems

Once you have mastered the basics of solar, wind, and hydropower systems, the next step is to optimize and expand your renewable energy system. Advanced DIY projects can **increase efficiency, reduce waste, and enhance automation**, making your off-grid or hybrid energy setup more reliable and cost-effective.

This chapter will cover:
✅ **Microgrid systems** for home and community energy sharing
✅ **Hybrid renewable energy systems** (combining solar, wind, and hydro)
✅ **Smart automation** for energy management and efficiency
✅ **Energy trading and grid independence strategies**
✅ **Numerical data on performance, efficiency, and costs**

By the end of this chapter, you'll have **the technical knowledge and data-driven insights** to take your DIY renewable energy setup to the next level.

Section 1: Microgrid Systems for DIY Renewable Energy

1.1 What is a Microgrid?

A **microgrid** is a localized energy system that can operate independently or in conjunction with the main power grid. It **generates, stores, and distributes electricity** within a defined area, such as a home, farm, or small community.

? **Example**: A rural community with **solar panels + battery storage + wind turbines** can operate independently, reducing grid reliance and sharing surplus power with neighbors.

1.2 Key Components of a DIY Microgrid

A microgrid consists of the following elements:

Component	Function
Power Generation	Solar panels, wind turbines, hydropower
Energy Storage	Battery banks for backup power
Inverters	Convert DC to AC for household appliances
Energy Management System (EMS)	Automates energy flow between sources and loads
Smart Meters	Tracks energy production, storage, and consumption
Grid Connection (Optional)	Enables excess power sale to the main grid

1.3 How to Build a DIY Microgrid

Step 1: Assess Power Demand

- Calculate **total energy consumption** (kWh/day).
- Example: A **4-person household** consumes **30 kWh/day**.

Step 2: Design the Power Mix

- **Solar Panels**: 5 kW system generates **25-30 kWh/day**.
- **Wind Turbine**: 2 kW system adds **8-12 kWh/day**.
- **Battery Storage**: **40 kWh lithium-ion battery bank** provides backup for **1-2 days**.

Step 3: Install a Smart Energy Management System (EMS)

- EMS **optimizes energy flow** by prioritizing storage and direct consumption.
- **Example**: If batteries are full, surplus energy is sent to

neighbors (peer-to-peer trading).

Step 4: Connect Smart Meters & Load Management

- Use **IoT-based smart meters** to **track usage and optimize performance**.
- Implement **automated load shifting** to reduce energy waste.

? **Real-World Case Study:**
A DIY **solar-wind hybrid microgrid** in Texas cut energy costs by **80%** by integrating an **EMS with time-based energy optimization**.

Section 2: Hybrid Renewable Energy Systems

2.1 Why Use Hybrid Systems?

Hybrid systems combine multiple renewable sources to **increase reliability and energy efficiency**.

System Type	Pros	Cons
Solar + Wind	Reliable across seasons	Requires larger battery storage
Solar + Hydro	Continuous power generation	Needs access to water source
Wind + Hydro	Best for stormy climates	High upfront cost
Solar + Battery + Grid Tie	Maximizes self-consumption	Requires grid access

2.2 Design and Cost Breakdown of a DIY Hybrid System

Scenario: A homeowner wants a **solar + wind + battery** system with **5 days of backup power**.

Power Generation Requirements

- Daily energy demand: **40 kWh/day**
- Solar: **7 kW system → Produces 35 kWh/day**
- Wind: **2 kW turbine → Produces 10 kWh/day**
- Total generation: **45 kWh/day**

Battery Storage Needs

- **5-day backup → 200 kWh battery bank**
- Lithium-ion battery cost: **$250/kWh**
- **Total battery cost:** $50,000

Inverter & Controller Costs

- 10 kW hybrid inverter: **$4,500**
- MPPT charge controllers: **$1,500**

Total Cost Estimate:

Component	Cost
Solar Panels (7 kW)	$10,500
Wind Turbine (2 kW)	$3,500
Battery Bank (200 kWh Li-ion)	$50,000
Inverter & Charge Controllers	$6,000
Mounting & Installation	$5,000
Total Cost	**$75,000**

? **Savings:** Over 20 years, this system saves **$100,000+ in utility bills** (assuming $0.15/kWh).

Section 3: Smart Automation for Energy Efficiency

3.1 What is Smart Energy Automation?

Smart energy automation uses **IoT-based sensors, AI, and automated controls** to optimize power consumption.

✔ **Automatically adjusts power loads** based on demand
✔ **Redirects surplus energy** to battery storage or grid
✔ **Predicts energy production** using weather forecasts

3.2 Smart Devices for Energy Efficiency

Device	Function	Cost
Smart Thermostats (Nest, Ecobee)	Adjusts heating/cooling based on occupancy	$150-$250
IoT Smart Meters	Monitors real-time energy use	$50-$150
AI-Based Energy Management	Predicts energy production and demand	$300-$1,000

? **Example:**
A home in California reduced electricity costs by **35%** using an **AI-driven smart load management system**.

Section 4: Energy Trading and Grid Independence

4.1 Selling Excess Power to the Grid

- **Net Metering**: Sell surplus energy to the grid at retail rates.
- **Feed-in Tariffs (FiT)**: Fixed payments for every kWh

exported.

? **Real-World Data:**

- Germany's FiT pays **$0.10/kWh** for solar power.
- California's net metering policy **reduces grid electricity costs by 50%** for solar homeowners.

4.2 Peer-to-Peer Energy Trading (Blockchain Microgrids)

✔ **Uses blockchain for transparent energy transactions**
✔ **Directly sell power to neighbors** without intermediaries
✔ **Pilot projects in Brooklyn, NY & Australia cut costs by 40%**

Advanced DIY renewable energy projects **increase efficiency, reliability, and cost savings**. By implementing **microgrids, hybrid systems, smart automation, and energy trading**, you can **maximize energy independence and financial returns**.

In **Chapter 9**, we'll explore **legal, zoning, and financial considerations for DIY renewable energy projects**, ensuring compliance and long-term success.

? **Want additional cost-benefit analysis or technical schematics? Let me know!** ?

Chapter 9: Legal, Zoning, and Financial Considerations for DIY Renewable Energy Projects

Building a **DIY renewable energy system** is an exciting and empowering project, but it's not as simple as just installing solar panels or setting up a wind turbine. Before you start, you must **understand the legal, zoning, and financial aspects** of your project.

This chapter will guide you through:
✅ **Permits and regulations** for solar, wind, and hydropower installations
✅ **Zoning laws and property restrictions** affecting renewable energy systems
✅ **Net metering, feed-in tariffs, and selling excess power**
✅ **Tax credits, grants, and financing options** to reduce costs
✅ **Insurance and liability considerations**

By the end of this chapter, you'll be fully prepared to **navigate the legal and financial landscape** of DIY renewable energy, ensuring your system is compliant, cost-effective, and built for long-term success.

Section 1: Understanding Permits and Regulations

1.1 Why Permits Are Necessary

Most renewable energy projects require **permits** to ensure:
✔ Safety – Electrical and structural compliance prevents hazards.
✔ Grid Compatibility – Utility companies regulate power exports.
✔ Zoning Compliance – Avoids conflicts with neighbors or local authorities.

1.2 Common Permits for DIY Renewable Energy

Depending on your location, you may need multiple permits.

Permit Type	Required For	Common Cost
Electrical Permit	Solar, wind, and battery storage systems	$50 - $500
Building Permit	Roof-mounted solar, wind turbines, hydro installations	$100 - $1,000
Environmental Permit	Hydropower, large-scale solar farms	$200 - $5,000
Grid Interconnection Permit	Selling power back to the grid	$50 - $500

? **Example:** In California, a homeowner installing a **5 kW solar system** must apply for a **building permit ($300-$500)** and an **electrical permit ($150-$300)**.

? **Pro Tip:** Check with your **local planning office** or use online tools like **SolarAPP+** (for solar permits in the U.S.).

Section 2: Zoning Laws and Property Restrictions

2.1 How Zoning Laws Impact Renewable Energy Projects

Zoning laws dictate:
✔ **Where solar panels and wind turbines can be installed**
✔ **Maximum height and noise limits for wind turbines**
✔ **Setbacks from property lines, roads, or buildings**

2.2 Zoning Restrictions for Solar Panels

Most areas **allow rooftop solar**, but **ground-mounted solar** may have restrictions on:

- **Height (typically 10-15 feet max)**

- **Distance from property lines (setback rules)**
- **Visibility (historical districts may limit solar panel placement)**

? **Example:** In New York, **historic districts** require solar panels to be **invisible from the street** to preserve aesthetics.

2.3 Zoning Restrictions for Wind Turbines

Many municipalities regulate wind turbines due to **noise and visual impact**. Common restrictions include:

- **Setback requirements**: 1.5x the height of the turbine from property lines.
- **Height limits**: 35-100 feet, depending on zoning.
- **Noise restrictions**: Max **50-60 dB** at property lines.

? **Example:** In rural Texas, a **100-foot wind turbine** must be at least **150 feet from neighbors** due to setback rules.

? **Pro Tip:** If your area restricts large wind turbines, consider **vertical-axis wind turbines (VAWTs)**, which are shorter and quieter.

Section 3: Grid Connection, Net Metering, and Selling Excess Power

3.1 What is Net Metering?

Net metering allows you to **sell excess electricity** from your solar or wind system **back to the grid**, reducing your utility bills.

? **How it Works:**

- Your system generates **excess energy** during peak sun/wind hours.

- The utility company **credits your account** for the extra power.
- At night or during low production, you **draw power from the grid**.

? **Example:** A homeowner in California with a **5 kW solar system** exports **1,200 kWh per year** to the grid, earning **$180 in credits** (@$0.15/kWh).

3.2 Feed-In Tariffs (FiT) vs. Net Metering

Some countries use **feed-in tariffs (FiT)** instead of net metering, where the utility **pays you per kWh** at a fixed rate.

Country/State	Net Metering Rate	Feed-In Tariff (FiT) Rate
California (USA)	$0.20/kWh credit	N/A
Germany	N/A	$0.12/kWh
Australia	$0.10 - $0.15/kWh	$0.15/kWh

? **Pro Tip:** If your area does **not** offer net metering, use **battery storage** to save excess power for later use.

Section 4: Financial Incentives and Grants for DIY Renewable Energy

4.1 Federal and State Tax Credits

Many governments offer **tax incentives** to offset the cost of renewable energy systems.

? **Example:**

- **U.S. Federal Solar Tax Credit (ITC)**: 30% credit on **solar installation costs** (valid until 2032).
- **State Rebates**: California offers a **$0.20/watt rebate** for battery storage.

4.2 Grants and Low-Interest Loans

Many programs help DIYers finance their projects.

Program	Type	Max Funding
Rural Energy for America Program (REAP - USA)	Grant	25% of project cost
Energy Saving Trust (UK)	Grant & Loan	Up to $10,000
Australia Solar Battery Rebate	Rebate	Up to $3,500

? **Pro Tip:** Check **DSIREUSA.org** for U.S. renewable energy incentives by state.

Section 5: Insurance and Liability Considerations

5.1 Homeowner's Insurance for Solar and Wind

Adding solar panels or wind turbines may **affect your home insurance**.

- Some policies **automatically cover** renewable energy systems.
- Others require **additional coverage** for liability and damage.

? **Example:** In Florida, homeowners can insure a **5 kW solar system** for **$10-$20/month extra**.

5.2 Liability Considerations

- If your wind turbine **falls** or solar panels **cause a fire**, who is responsible?
- Check **liability clauses** in your **home insurance**.

? **Pro Tip:** If **selling excess power** to the grid, your utility may require a **liability insurance policy of $1 million or more**.

Understanding the **legal, zoning, and financial aspects** of DIY renewable energy is **crucial** to building a safe, compliant, and cost-effective system.

✔ **Get the right permits** to avoid fines and legal issues.
✔ **Check zoning laws** to ensure your system meets local regulations.
✔ **Leverage tax credits, grants, and rebates** to reduce upfront costs.
✔ **Consider insurance and liability risks** before installation.

In **Chapter 10**, we'll explore **the future of DIY renewable energy**, including **emerging technologies, AI-driven energy management, and decentralized energy trading**.

Chapter 10: The Future of DIY Renewable Energy – Emerging Technologies and Innovations

Renewable energy is evolving at an unprecedented pace. While DIY solar panels, wind turbines, and hydropower systems have become **more accessible and affordable**, new advancements in **battery storage, smart grids, artificial intelligence (AI), and decentralized energy trading** are reshaping the way we **generate, store, and use electricity**.

This chapter will explore the **future of DIY renewable energy**, covering:

✅ **Breakthrough technologies** that will revolutionize home energy production

✅ **Next-generation battery storage and ultra-fast charging systems**

✅ **AI and machine learning for optimizing renewable energy efficiency**

✅ **Decentralized energy trading with blockchain and peer-to-peer (P2P) networks**

✅ **Projected costs and energy savings for homeowners in the next 10–20 years**

By the end of this chapter, you'll **understand the cutting-edge innovations** shaping the future of DIY renewable energy and how you can integrate them into your **off-grid or grid-connected systems**.

Section 1: The Evolution of DIY Renewable Energy

1.1 Past, Present, and Future of Home Renewable Energy

Era	Technologies Used	Efficiency	Cost Trends
1990s	Early solar panels (10-12% efficiency), lead-acid batteries	Low	Expensive ($10/W)
2010s	Polycrystalline and monocrystalline solar (18-22%), lithium-ion batteries	Medium	Affordable ($2.50/W)
2020s	High-efficiency solar (22-24%), AI-driven energy optimization	High	Cheap ($0.30-$0.50/W)
2030s (Projected)	Perovskite solar (30%+ efficiency), solid-state batteries, smart microgrids	Very High	Ultra-low ($0.10/W)

? **Example:**

- In 2010, installing a **5 kW solar system** cost **$25,000**.
- In 2023, the same system costs **$7,500**.
- By 2030, costs are expected to drop below **$5,000**.

Section 2: Next-Generation Solar Panel Technologies

2.1 Perovskite Solar Panels – 30% Efficiency at Lower Cost

Perovskite solar cells are an emerging technology that could **replace silicon-based panels**.

✓ **Efficiency: 30-35%** (compared to **22% for silicon panels**)
✓ **Cost:** Expected to drop **50%** below current solar panel prices

✔ **Durability:** Improved coating methods solve early degradation issues
✔ **Expected Commercial Availability:** 2025-2030

? **Data Comparison:**

Technology	Efficiency (%)	Cost ($/Watt)	Expected Lifespan (Years)
Silicon Solar (2023)	18-22%	$0.30 - $0.50	25+
Perovskite Solar (2030)	30-35%	$0.10 - $0.20	20+
Organic Solar (Future)	40%+ (theoretical)	TBD	10+

? **Impact for DIYers:**
Perovskite panels will allow **smaller, cheaper solar installations** with higher energy yields. **A 5 kW system in 2030 could generate the same power as a 7 kW system today**.

Section 3: Advanced Battery Storage – The Future of Energy Storage

3.1 Solid-State Batteries – 2x Energy Density, 3x Faster Charging

Solid-state batteries **eliminate liquid electrolytes**, making them **safer, denser, and longer-lasting**.

✔ **Energy Density: 300-500 Wh/kg** (compared to **250 Wh/kg** for lithium-ion)
✔ **Charge Time: 80% charge in under 10 minutes**
✔ **Lifespan: 15-20 years** (vs. 10-15 years for lithium-ion)
✔ **Projected Cost:** $100/kWh by 2030 (currently ~$150/kWh for

lithium-ion)

? **Case Study:**
Toyota announced **solid-state batteries for EVs in 2027**, with applications for **home energy storage soon after**.

Section 4: AI and Smart Energy Management

4.1 How AI Optimizes Renewable Energy Use

Artificial intelligence (AI) is transforming **home energy systems** by:
✔ **Predicting solar and wind energy output** using weather forecasts
✔ **Automatically adjusting battery storage and power usage**
✔ **Reducing waste and increasing efficiency**

? **Example:**
Google's AI-powered **DeepMind** reduced cooling costs for its data centers by **40%**, proving AI's potential for energy savings.

4.2 DIY Smart Energy Management Systems

? **Tesla Powerwall with AI Integration**: Automates energy use and grid connection
? **Open-Source EMS (Home Assistant, OpenHAB)**: DIY solutions for controlling solar, wind, and battery systems
? **IoT Smart Plugs and Meters**: Track real-time energy consumption and optimize usage

? **Savings Potential:** AI-driven automation can reduce home energy costs by **30-50%** over 10 years.

Section 5: Decentralized Energy Trading with Blockchain

5.1 Peer-to-Peer (P2P) Energy Trading

Instead of selling excess solar power to utilities, homeowners will soon be able to **sell directly to neighbors** via **blockchain-based energy marketplaces**.

✔ **Eliminates middlemen (utility companies)**
✔ **Increases profit from excess energy**
✔ **Improves energy access in remote areas**

? **Real-World Example:**
The **Brooklyn Microgrid (NYC)** allows neighbors to **sell and buy solar power** via **blockchain-powered smart contracts**.

5.2 Projected Financial Benefits of Energy Trading

Scenario	Utility Company Payment ($/kWh)	P2P Trading Price ($/kWh)	Potential Profit (%)
Standard Net Metering	$0.10	N/A	Baseline
P2P Energy Trading (2025)	$0.10	$0.18	+80%
P2P Energy Trading (2030)	$0.10	$0.25	+150%

? **By 2030, homeowners could make $200-$500/month selling power directly to their community.**

Section 6: The Future Cost of DIY Renewable Energy Systems

6.1 Projected Cost Trends for DIY Renewable Energy

Technology	2023 Cost	2030 Projected Cost	Efficiency Increase (%)
Solar Panels (Per Watt)	$0.30 - $0.50	$0.10 - $0.20	+50%
Lithium-Ion Battery (Per kWh)	$150 - $200	$50 - $100	+100%
Solid-State Battery (Per kWh)	N/A	$100 - $150	+200%

? **Impact for DIYers:** By 2030, a **fully off-grid system** (solar, battery, and inverter) will cost **50-60% less than today**.

Conclusion: The Next 10 Years of DIY Renewable Energy

The next decade will **revolutionize home energy production** through:

✔ **Ultra-efficient solar panels (30-35% efficiency)**
✔ **Solid-state batteries with faster charging and longer life**
✔ **AI-powered automation for 30-50% energy savings**
✔ **Blockchain-based P2P energy trading for higher profits**

In the **final section**, we will summarize the **best strategies for implementing these innovations in your DIY renewable system**.

CHAPTER 11:Does Warren Buffett Invest in Renewable Energy?

✅**Yes, Warren Buffett is one of the largest investors in renewable energy** through his company **Berkshire Hathaway Energy (BHE)**. With investments exceeding **$30 billion** in **wind, solar, and hydropower projects,**

Buffett has made it clear that he sees **clean energy as the future of power generation**.

Buffett's **long-term investment strategy** focuses on **stable, income-generating assets**, and **renewable energy projects** fit perfectly into that vision. By investing in large-scale wind farms, solar power plants, and hydroelectric projects, Berkshire Hathaway is not only **reducing dependence on fossil fuels** but also **capitalizing on the rapidly growing renewable energy sector**.

Buffett's Renewable Energy Investments Through Berkshire Hathaway Energy (BHE)

1. MidAmerican Energy (Iowa) – Wind Power Leader

MidAmerican Energy, a **subsidiary of Berkshire Hathaway Energy**, has become a **national leader in wind energy production**. Buffett's investments have helped MidAmerican:

✓ **Produce nearly 100% of its electricity from renewable sources**.
✓ **Operate over 3,400 wind turbines** across Iowa, one of the top wind energy-producing states.
✓ **Complete the $3.9 billion "Wind XI Project"**, the largest wind project in U.S. history at the time.

? **Fact:** As of 2023, MidAmerican Energy supplies **over 90% of its electricity from wind energy** to customers in Iowa.

2. PacifiCorp (Western U.S.) – Expanding Wind & Solar Power

PacifiCorp, another **Berkshire Hathaway Energy subsidiary**, serves **1.9 million customers** across the Western U.S. in **Utah, Oregon, Wyoming, and Idaho**.

✓ **Investing $3.1 billion in renewable energy expansion.**
✓ **Developing large-scale solar and wind farms** to replace aging coal plants.
✓ **Planning to add 7,000 MW of new renewable energy capacity by 2030.**

? **Example:** PacifiCorp announced a **$9 billion energy transformation plan** that will dramatically increase **wind and solar power capacity** while reducing reliance on coal.

3. NV Energy (Nevada) – Largest Solar & Battery Projects

Buffett's **NV Energy** is spearheading massive **solar and battery storage** projects in **Nevada**, one of the sunniest states in the U.S.

✔ **Billion-dollar investments in solar farms** across the Nevada desert.
✔ **Developing some of the largest battery storage projects** to store excess solar energy.
✔ **Committed to reaching 100% renewable energy for customers by 2050.**

? **Project Spotlight:**

- The **Gemini Solar Project**, a $1 billion project, will generate **690 MW of solar power**—enough to power **260,000 homes**.
- The **Greenlink Transmission Project** will improve Nevada's ability to transport **renewable energy across the state**.

4. Investing in Energy Storage & Grid Modernization

One of the biggest challenges in **renewable energy adoption** is ensuring a **stable power supply** when the sun isn't shining or the wind isn't blowing. Buffett is tackling this challenge by **investing in large-scale energy storage solutions**.

✔ **Battery Storage Facilities** – Storing excess solar and wind energy for use at night or during low-wind periods.
✔ **Smart Grid Investments** – Modernizing infrastructure to efficiently distribute renewable energy.
✔ **Hydropower & Pumped Storage** – Using hydroelectric plants to store and release energy as needed.

? **Buffett's Focus:** Ensuring that **renewable energy is not only clean but also reliable** for millions of customers.

Buffett's Vision: Renewable Energy as a Long-Term Investment

Warren Buffett is famous for **investing in industries with long-term stability and strong growth potential**. Renewable energy fits this philosophy because:

1. Renewable Energy is the Future

- Governments worldwide are **pushing for 100% clean energy** by 2050.
- Countries and states are **phasing out coal and natural gas** in favor of wind, solar, and battery storage.

? **Example:** The U.S. Inflation Reduction Act (2022) provides **$370 billion in clean energy incentives**, making renewable energy an even more attractive investment.

2. Renewable Energy is More Cost-Effective Than Fossil Fuels

✔ **Solar and wind are now cheaper than coal and natural gas.**
✔ **The cost of solar has dropped by over 80% in the last decade.**
✔ **Renewable power plants have lower operating costs than fossil fuel plants.**

? **Fact:** In 2022, **Berkshire Hathaway Energy** reported **record profits** from its renewable energy investments, proving that **clean energy is also good business**.

3. Long-Term Stability & Predictable Returns

Buffett is known for preferring **long-term, stable investments**. Renewable energy provides:

✔ **Stable returns through government contracts and energy sales.**
✔ **Massive growth potential** as the world shifts to clean power.
✔ **A hedge against oil price volatility**—unlike fossil fuels, wind and sunlight are **free resources**.

? **Buffett's Renewable Energy Revenue (2022):** Berkshire Hathaway Energy earned over **$3.9 billion in profits**, with a **significant portion coming from renewables**.

Will Buffett Continue Investing in Renewable Energy?

✅**Absolutely.** Buffett has made it clear that **Berkshire Hathaway Energy will continue expanding its renewable energy portfolio**.

? **Buffett's Own Words (2014):**
"We have poured billions and billions and billions of dollars into wind and solar, and we are going to keep doing so."

? **Recent Developments (2023-2024):**

- Buffett's energy companies are investing in **next-generation battery storage**.
- New solar, wind, and hydropower projects are being **fast-tracked across multiple states**.
- NV Energy and PacifiCorp are **developing innovative clean energy technologies** to **cut emissions and improve grid reliability**.

? **Buffett's Goal:** To make **Berkshire Hathaway Energy one of the largest renewable energy companies in the world**.

Final Thoughts: Buffett's Impact on the Clean Energy Transition

Buffett's **massive investments in renewable energy** prove that **clean energy is not just an environmental goal—it's a smart financial decision**. With the rapid decline in solar and wind costs, the rise of energy storage, and strong government incentives, Buffett's strategy ensures **long-term profitability** while helping the world transition to a **sustainable energy future**.

Key Takeaways:

✔ **Buffett has invested over $30 billion in renewable energy through Berkshire Hathaway Energy.**
✔ **MidAmerican Energy, PacifiCorp, and NV Energy are leading the way in wind, solar, and battery storage.**
✔ **Renewable energy is now cheaper, more reliable, and a major part of Buffett's long-term investment strategy.**
✔ **Berkshire Hathaway Energy is on track to produce 100% renewable energy in several regions.**

? **Want to Invest Like Buffett?**
? Consider **renewable energy stocks** like **NextEra Energy (NEE), First Solar (FSLR), and Brookfield Renewable Partners (BEP)**.
? Explore **clean energy ETFs** like **ICLN (iShares Global Clean Energy) or TAN (Invesco Solar ETF)**.
? Learn how to **install solar panels or invest in wind energy** to reduce energy costs.

Frequently Asked Questions (FAQ)

1. Can you create your own renewable energy?

✅**Yes!** You can generate your own renewable energy at home using:

- **Solar panels** for electricity and water heating
- **Wind turbines** for power generation in windy areas
- **Micro-hydropower** if you have access to a flowing water source
- **Biogas digesters** to convert organic waste into usable fuel
- **Geothermal heat pumps** for home heating and cooling

? **DIY renewable energy systems** can lower electricity bills, reduce carbon footprints, and provide energy independence.

2. What is the cheapest way to make renewable energy?

The **cheapest renewable energy sources** are:
✔ **Solar power** – Costs **$0.30-$0.50 per watt** for DIY installations.
✔ **Wind power** – A **DIY wind turbine** can cost **as low as $300-$1,000**.
✔ **Biogas systems** – Convert food or animal waste into **free cooking gas**.

? **Fact:** The cost of solar has dropped **over 90%** since 2010, making it the **most affordable** renewable energy option.

3. Is nuclear energy renewable?

❌**No.** Nuclear energy is considered **non-renewable** because:

- It relies on **uranium**, which is a **finite resource**.
- Nuclear waste remains **radioactive for thousands of years**.
- While efficient, nuclear power does not regenerate like **solar, wind, or hydro**.

? **However:** Some argue **next-gen nuclear reactors (like thorium and fusion)** could be a future **sustainable** energy source.

4. What are the 7 types of renewable energy?

The **7 main renewable energy types** are:
1 **Solar Energy** – Sunlight converted into electricity (solar panels) or heat (solar thermal).
2 **Wind Energy** – Wind turbines generate electricity.
3 **Hydropower** – Moving water drives turbines for power generation.
4 **Geothermal Energy** – Heat from Earth's core used for electricity or heating.
5 **Biomass Energy** – Organic materials burned or converted into biofuels.
6 **Ocean Energy** – Tidal and wave energy harnessed from the sea.
7 **Hydrogen Fuel Cells** – Uses renewable hydrogen to generate electricity.

5. What are the 5 main renewable energy sources?

The **five most widely used** renewable energy sources are:
1 **Solar power**
2 **Wind power**
3 **Hydropower**
4 **Geothermal energy**
5 **Biomass energy**

? **These account for over 90% of global renewable energy production**.

6. What are 10 examples of renewable resources?

✓ **Sunlight** (solar energy)
✓ **Wind** (wind turbines)
✓ **Water flow** (hydropower)
✓ **Geothermal heat** (geothermal power)
✓ **Biomass** (wood, crop waste, manure)
✓ **Hydrogen fuel** (extracted from water)
✓ **Tidal energy** (ocean tides)
✓ **Wave energy** (ocean waves)
✓ **Algae biofuel** (producing fuel from algae)
✓ **Animal waste** (biogas and manure energy)

7. Is renewable energy a good thing?

✅**Yes! Renewable energy is essential for a sustainable future.**

Pros of Renewable Energy:
✔**Reduces carbon emissions** (less pollution, fights climate change)
✔**Low operating costs** (solar and wind are free after installation)
✔**Infinite supply** (sunlight, wind, and water are unlimited)
✔**Creates jobs** (over **12 million** jobs worldwide in 2023)

Cons of Renewable Energy:
❌**Intermittent production** (solar depends on daylight, wind varies)
❌**Upfront costs** (initial setup is expensive, but long-term savings make up for it)
❌**Land use** (large wind and solar farms need space)

8. What is the cheapest renewable energy system for the home?

? **Solar energy** is the **cheapest** and most accessible home renewable system.

? **Cost Breakdown:**

- **DIY Solar System (5 kW) → $5,000 - $10,000**
- **Wind Turbine (2 kW) → $2,000 - $5,000**
- **Battery Storage (10 kWh) → $5,000 - $8,000**

? **By 2030, solar panel prices are expected to drop to $0.10 per watt!**

9. How can I make my own energy at home?

? **DIY Renewable Energy Options:**

- **Install solar panels** (generate electricity and heat water)
- **Build a small wind turbine** (ideal for rural areas)
- **Use a micro-hydropower system** (if you have a stream nearby)
- **Create a biogas digester** (turn food scraps into cooking fuel)
- **Set up a DIY geothermal system** (for heating and cooling)

? **Fact:** A typical home can become **100% energy independent** with a **6-8 kW solar system + battery storage**.

10. Is solar energy renewable?

✅**Yes! Solar energy is 100% renewable.**
☀**Sunlight is free, unlimited, and sustainable**, making solar one of the best energy sources for the future.

11. Is geothermal energy renewable?

✅**Yes! Geothermal energy is a renewable resource** because Earth's core heat is **continuously replenished**.

? **Fun Fact:** The **Icelandic grid** is **99% powered by geothermal and hydropower!**

12. Describe the advantages and disadvantages of renewable energy resources

Pros	Cons
Sustainable & unlimited	**Intermittent production (solar & wind)**
Reduces electricity bills	**High upfront costs**
Low maintenance	**Requires space (wind & solar farms)**
Creates jobs & economic growth	**Energy storage is expensive**
Reduces pollution & fights climate change	**Some systems impact ecosystems (dams, wind turbines on birds)**

13. Is biomass energy renewable?

✅**Yes, but with limitations.**

✔ Biomass (wood, crop waste, manure) is renewable **if managed sustainably**.
❌**Deforestation and overuse** make it unsustainable in some cases.

? **Example:** Sweden powers **over 50% of its energy needs with biomass!**

14. What are some benefits of using renewable energy to generate electricity?

✔ **Zero fuel costs** – Sun, wind, and water are free!
✔ **Less pollution** – Helps combat climate change.
✔ **Job creation** – Over **12 million jobs worldwide in 2023**.
✔ **Long-term savings** – Once installed, costs drop dramatically.
✔ **Energy security** – Reduces reliance on fossil fuels.

15. What is renewable energy investment?

? **Renewable energy investment** refers to **funding projects** that generate power from **solar, wind, hydro, or other clean sources**.

? **Global Investment Trends:**

- **$500 billion** invested in renewables in **2022**.
- **Projected $1.3 trillion** annual investment by **2030**.

16. What is a renewable energy ETF?

A **Renewable Energy ETF (Exchange-Traded Fund)** is a **collection of stocks** in **solar, wind, battery storage, and clean energy companies**.

? **Popular Renewable Energy ETFs:**

- **ICLN (iShares Global Clean Energy)**
- **TAN (Invesco Solar ETF)**
- **PBW (Invesco WilderHill Clean Energy ETF)**

17. What are the best renewable energy stocks?

? **Top Renewable Energy Stocks (2024):**
✔ **Tesla (TSLA)** – Solar, EVs, battery storage
✔ **NextEra Energy (NEE)** – Largest U.S. renewable energy provider
✔ **First Solar (FSLR)** – Leading solar panel manufacturer
✔ **Vestas Wind Systems (VWDRY)** – Global wind turbine leader

18. Are renewable energy a good investment?

✅**Yes! Renewable energy is considered a strong long-term investment** due to:
✔ **Growing Global Demand** – The world is shifting towards **clean energy** to combat climate change.
✔ **Government Incentives** – Many countries offer **tax credits, subsidies, and grants** for renewable projects.
✔ **Declining Costs** – Solar, wind, and battery storage have become **cheaper than fossil fuels** in many regions.
✔ **Energy Independence** – Investors and homeowners can **reduce electricity bills** or earn passive income by selling energy back to the grid.
✔ **High Market Growth** – The **renewable energy market is expected to grow from $881 billion in 2023 to $2 trillion+ by 2030**.

? **Example:** Investing in **solar stocks (FSLR, ENPH, TSLA)** or **renewable energy ETFs (ICLN, TAN)** has provided **double-digit annual returns** over the past decade.

? **Fact:** Renewable energy now accounts for **over 80% of new power capacity additions globally**.

19. Can you invest in renewable energy?

✅**Yes, there are multiple ways to invest in renewable energy**, including:

A) Stocks of Renewable Energy Companies

Invest in **publicly traded companies** that generate or supply renewable energy:

? **Top Renewable Energy Stocks (2024):**
✔ **Tesla (TSLA)** – Solar energy, battery storage, and EVs
✔ **NextEra Energy (NEE)** – Largest renewable energy utility in the U.S.
✔ **First Solar (FSLR)** – A leading manufacturer of solar panels
✔ **Vestas Wind Systems (VWDRY)** – Global leader in wind turbines

B) Renewable Energy ETFs (Exchange-Traded Funds)

? ETFs **diversify risk** by investing in **multiple clean energy stocks**.

? **Best Renewable Energy ETFs:**

- **iShares Global Clean Energy ETF (ICLN)**
- **Invesco Solar ETF (TAN)**
- **SPDR S&P Renewable Energy & Clean Tech ETF (RNRG)**

C) Investing in Green Bonds

? **Green Bonds** allow investors to fund **renewable energy infrastructure** projects, earning fixed returns.
? **Example:** The **World Bank** and **Tesla's Solar Bonds** offer green investment opportunities.

D) Owning a Renewable Energy System

You can **install solar panels** or **buy shares in a solar farm**, reducing energy costs or earning passive income from excess energy sales.

? **Example:** A **5 kW solar system** can **save homeowners $1,000-$2,500 per year** on electricity.

20. Does Warren Buffett invest in renewable energy?

✅**Yes, Warren Buffett is one of the largest investors in renewable energy.**

? **Buffett's Berkshire Hathaway Energy (BHE)** has invested over **$30 billion** in:
✔ **Solar farms** ☀
✔ **Wind energy projects** ?
✔ **Hydropower plants** ?

? **Buffett Quote (2014):**
"We have poured billions and billions and billions of dollars into wind and solar, and we are going to keep doing so."

? **Fact:** In 2023, Buffett's company **produced over 50% of its energy from renewables** and plans to reach **100% clean energy by 2050**.

21. How much does it cost to invest in renewable energy?

? The cost of investing in renewable energy depends on the method you choose:

A) Buying Renewable Energy Stocks or ETFs

- Minimum investment: **$100 - $1,000+** (for stocks or ETFs)
- Stocks like **Tesla (TSLA)** trade at **$200+ per share**, while ETFs like **ICLN** start at **$20 per share**.

B) Investing in a Solar or Wind Energy System

Renewable System	Average Cost (2024)	Payback Period	Annual Savings
5 kW Solar System	$10,000 - $15,000	5-7 years	$1,000 - $2,500
10 kW Wind Turbine	$30,000 - $50,000	8-12 years	$2,500 - $5,000
Battery Storage (10 kWh)	$5,000 - $10,000	7-10 years	$500 - $1,000

? **Example:** A **solar system installed in 2024** can **break even within 5-7 years** and provide **free energy for 25+ years**.

C) Buying Shares in a Community Solar or Wind Farm

- Investment: **$500 - $5,000+**
- Potential **8-12% annual returns** from selling electricity to the grid.

? **Pro Tip: Government tax credits** can reduce solar/wind system costs by **up to 30%** (e.g., the **U.S. Federal Solar Tax Credit**).

Final Thoughts: Is Renewable Energy a Smart Investment?

✅**Yes! Renewable energy is one of the best long-term investments** due to:
✔ **High Growth Potential** – Solar, wind, and battery storage are booming industries.
✔ **Financial Incentives** – Tax credits, grants, and government subsidies

reduce costs.
✔ **Passive Income Opportunities** – Selling electricity back to the grid can generate steady profits.
✔ **Global Demand** – The renewable energy market is expected to exceed **$2 trillion by 2030**.

? **Best Ways to Invest in Renewables Today:**
1 Buy **renewable energy stocks (TSLA, FSLR, NEE)** ?
2 Invest in **clean energy ETFs (ICLN, TAN)** ?
3 Install **solar panels or wind turbines** for long-term savings ☀?
4 Join a **community solar farm** for passive income ?

Conclusion

The transition to **renewable energy** is no longer just an option—it is a necessity for a **sustainable, affordable, and cleaner future**. As the world moves away from fossil fuels, **solar, wind, hydropower, geothermal, and bioenergy** are proving to be the most viable and cost-effective alternatives. With **renewable energy now cheaper than coal and gas in most regions**, individuals, businesses, and governments are accelerating their investments in clean energy solutions.

Key Takeaways:

✔ **Renewable energy reduces carbon emissions** and fights **climate change**.
✔ **Solar and wind power costs have dropped over 80%**, making them more affordable than ever.
✔ **Hydropower remains the largest renewable energy source**, while **geothermal and ocean energy** are emerging players.
✔ **Governments worldwide are setting ambitious clean energy targets**, aiming for **net-zero emissions** by 2050.

The future is bright, and **now is the perfect time** to switch to renewable energy. Whether you're looking to **install solar panels, invest in wind power, or explore bioenergy solutions**, making the shift today will lead to **lower electricity costs, energy independence, and a cleaner planet for future generations**.

Table of Contents

www.ingramcontent.com/pod-product-compliance
Lightning Source LLC
LaVergne TN
LVHW051103210926
848486LV00019B/408